ENERGY SECTOR STANDARD
OF THE PEOPLE'S REPUBLIC OF CHINA
中华人民共和国能源行业标准

Technical Specification for Flow Monitoring and Measuring of Hydraulic-Mechanical Auxiliary System in Hydroelectric Power Plants

水力发电厂水力机械辅助系统
流量监视测量技术规程

NB/T 10144-2019

Chief Development Department: China Renewable Energy Engineering Institute
Approval Department: National Energy Administration of the People's Republic of China
Implementation Date: October 1, 2019

China Water & Power Press

Beijing 2024

All rights reserved. No part of this publication may be reproduced, stored in a retrieval system, or transmitted in any form or by any means—electronic, mechanical, photocopying, recording or otherwise, without prior written permission of the publisher.

图书在版编目（CIP）数据

水力发电厂水力机械辅助系统流量监视测量技术规程：NB/T 10144-2019 = Technical Specification for Flow Monitoring and Measuring of Hydraulic-Mechanical Auxiliary System in Hydroelectric Power Plants (NB/T 10144-2019)：英文 / 国家能源局发布. -- 北京：中国水利水电出版社，2024. 10. -- ISBN 978-7-5226-2761-8

Ⅰ．TK7-65

中国国家版本馆CIP数据核字第2024HF4169号

ENERGY SECTOR STANDARD
OF THE PEOPLE'S REPUBLIC OF CHINA
中华人民共和国能源行业标准

Technical Specification for Flow Monitoring and Measuring of
Hydraulic-Mechanical Auxiliary System in Hydroelectric Power Plants
水力发电厂水力机械辅助系统流量监视测量技术规程
NB/T 10144-2019
（英文版）

Issued by National Energy Administration of the People's Republic of China
国家能源局　发布
Translation organized by China Renewable Energy Engineering Institute
水电水利规划设计总院　组织翻译
Published by China Water & Power Press
中国水利水电出版社　出版发行
　　Tel: (+ 86 10) 68545888　68545874
　　sales@mwr.gov.cn
　　Account name: China Water & Power Press
　　Address: No.1, Yuyuantan Nanlu, Haidian District, Beijing 100038, China
　　http: //www.waterpub.com.cn
中国水利水电出版社微机排版中心　排版
北京中献拓方科技发展有限公司　印刷
184mm×260mm　16开本　1.5印张　47千字
2024年10月第1版　2024年10月第1次印刷

Price（定价）：￥240.00

Introduction

This English version is one of China's energy sector standard series in English. Its translation was organized by China Renewable Energy Engineering Institute authorized by National Energy Administration of the People's Republic of China in compliance with relevant procedures and stipulations. This English version was issued by National Energy Administration of the People's Republic of China in Announcement [2023] No. 5 dated October 11, 2023.

This version was translated from the Chinese Standard NB/T 10144-2019, *Technical Specification for Flow Monitoring and Measuring of Hydraulic-Mechanical Auxiliary System in Hydroelectric Power Plants*, published by China Water & Power Press. The copyright is reserved by National Energy Administration of the People's Republic of China. In the event of any discrepancy in the implementation, the Chinese version shall prevail.

Many thanks go to the staff from the relevant standard development organizations and those who have provided generous assistance in the translation and review process.

For further improvement of the English version, any comments and suggestions are welcome and should be addressed to:

China Renewable Energy Engineering Institute
No. 2 Beixiaojie, Liupukang, Xicheng District, Beijing 100120, China
Website: www.creei.cn

Translating organization:

POWERCHINA Beijing Engineering Corporation Limited

Translating staff:

HAN Dongbin	WANG Yuhui	ZHANG Haosheng	GUO Xibing
ZHANG Languo	LIU Jiandong	ZHOU Zhenzhong	YI Zhongyou
ZHENG Dongfei			

Review panel members:

LI Zhongjie	POWERCHINA Northwest Engineering Corporation Limited
QIE Chunsheng	Senior English Translator
YAN Wenjun	Army Academy of Armored Forces, PLA
WANG Yuchuan	Northwest A&F University

CHEN Lei	POWERCHINA Zhongnan Engineering Corporation Limited
LIU Guofeng	POWERCHINA Northwest Engineering Corporation Limited
LI Kejia	POWERCHINA Northwest Engineering Corporation Limited
ZONG Wanbo	China Renewable Energy Engineering Institute
LI Shisheng	China Renewable Energy Engineering Institute

National Energy Administration of the People's Republic of China

翻译出版说明

本译本为国家能源局委托水电水利规划设计总院按照有关程序和规定，统一组织翻译的能源行业标准英文版系列译本之一。2023年10月11日，国家能源局以2023年第5号公告予以公布。

本译本是根据中国水利水电出版社出版的《水力发电厂水力机械辅助系统流量监视测量技术规程》NB/T 10144—2019 翻译的，著作权归国家能源局所有。在使用过程中，如出现异议，以中文版为准。

本译本在翻译和审核过程中，本标准编制单位及编制组有关成员给予了积极协助。

为不断提高本译本的质量，欢迎使用者提出意见和建议，并反馈给水电水利规划设计总院。

地址：北京市西城区六铺炕北小街2号
邮编：100120
网址：www.creei.cn

本译本翻译单位：中国电建集团北京勘测设计研究院有限公司

本译本翻译人员： 韩东斌　王雨会　张昊晟　郭喜兵
　　　　　　　　　张蓝国　刘建栋　周振忠　易忠有
　　　　　　　　　郑冬飞

本译本审核人员：

李仲杰　中国电建集团西北勘测设计研究院有限公司
郄春生　英语高级翻译
闫文军　中国人民解放军陆军装甲兵学院
王玉川　西北农林科技大学
陈　蕾　中国电建集团中南勘测设计研究院有限公司
刘国峰　中国电建集团西北勘测设计研究院有限公司
李可佳　中国电建集团西北勘测设计研究院有限公司
宗万波　水电水利规划设计总院
李仕胜　水电水利规划设计总院

国家能源局

Announcement of National Energy Administration of the People's Republic of China [2019] No. 4

National Energy Administration of the People's Republic of China has approved and issued 297 sector standards such as *Code for Electrical Design of Photovoltaic Power Projects*, including 105 energy standards (NB), 168 electric power standards (DL), and 24 petrochemical standards (NB/SH).

Attachment: Directory of Sector Standards

National Energy Administration of the People's Republic of China

June 4, 2019

Attachment:

Directory of Sector Standards

Serial number	Standard No.	Title	Replaced standard No.	Adopted international standard No.	Approval date	Implementation date
...						
17	NB/T 10144-2019	Technical Specification for Flow Monitoring and Measuring of Hydraulic-Mechanical Auxiliary System in Hydroelectric Power Plant			2019-06-04	2019-10-01
...						

Foreword

According to the requirements of Document GNKJ [2015] No. 12 issued by National Energy Administration of the People's Republic of China, "Notice on Releasing the Development and Revision Plan of the Second Batch of Energy Sector Standards in 2014", and after extensive investigation and research, summarization of practical experience, consultation of relevant Chinese standards, and wide solicitation of opinions, the drafting group has prepared this specification.

The main technical contents of this specification include: general provisions, terms, general requirements of flow monitoring and measuring, selection of monitoring points, selection of flow components, and layout and installation requirements of pipes.

National Energy Administration of the People's Republic of China is in charge of the administration of this specification. China Renewable Energy Engineering Institute has proposed this specification and is responsible for its routine management. Energy Sector Standardization Technical Committee on Hydropower Hydraulic Machinery is responsible for the explanation of specific technical contents. Comments and suggestions in the implementation of this specification should be addressed to:

China Renewable Energy Engineering Institute
No. 2 Beixiaojie, Liupukang, Xicheng District, Beijing 100120, China

Chief development organization:

POWERCHINA Beijing Engineering Corporation Limited

Participating development organizations:

Beijing Wanruida Monitoring Technology Co.

China Three Gorges Corporation

Chief drafting staff:

GOU Dongming	YI Zhongyou	ZHOU Zhenzhong	ZHENG Dongfei
LIU Wanjing	LI Wenxue	LI Shihong	WANG Yuhui
GUO Xibing	ZHANG Haosheng	ZHU Rusha	HUANG Yu
LIU Gongmei	YAO Zengmin		

Review panel members:

DAI Kangjun	YOU Chao	YUAN Lianjun	FU Guofeng

ZHANG Qiang	JIANG Dengyun	LI Yuebin	WU Saibo
WANG Jianhua	CHEN Dong	YANG Jiaming	QIN Daqing
SHAO Bao'an	PENG Shuojun	DU Gang	

Contents

1 General Provisions ································ 1
2 Terms ······································· 2
3 General Requirements of Flow Monitoring and Measuring ···································· 3
4 Selection of Monitoring Points ···················· 5
5 Selection of Flow Components ···················· 7
6 Layout and Installation Requirements of Pipes ········ 9
Explanation of Wording in This Specification ············ 10
List of Quoted Standards ···························· 11

1 General Provisions

1.0.1 This specification is formulated with a view to standardizing the design of flow monitoring and measuring of hydraulic-mechanical auxiliary system in hydroelectric power plants, to improve the safety and economy of the operation.

1.0.2 This specification is applicable to the flow monitoring and measuring of hydraulic-mechanical auxiliary system in hydroelectric power plants.

1.0.3 The flow monitoring and measuring of hydraulic-mechanical auxiliary system may be simplified for small-sized hydroelectric power plants.

1.0.4 This specification specifies the general requirements, selection of monitoring points, selection of flow components, and layout and installation requirements for flow monitoring and measuring of hydraulic-mechanical auxiliary system in hydroelectric power plants.

1.0.5 In addition to this specification, the flow monitoring and measuring of hydraulic-mechanical auxiliary system in hydroelectric power plants shall comply with other current relevant standards of China.

2 Terms

2.0.1 flow components

flow monitoring and measuring device such as flowmeter and flow switch

2.0.2 flowmeter

device for monitoring the flow rate of a fluid passing through a pipe. The flowmeters can be classified, by measuring method, into electromagnetic flowmeter, ultrasonic flowmeter, mechanical flowmeter, differential pressure flowmeter, etc.; and by installation method, into plug-in flowmeter, flange-connected flowmeter, etc.

2.0.3 flow switch

component for monitoring the magnitude of flow rate of a fluid passing through a pipe, which sends out a signal when the flow rate is higher or lower than the set value. The flow switches can be classified into thermal conductivity flow switch, mechanical flow switch, etc.

3 General Requirements of Flow Monitoring and Measuring

3.0.1 The monitoring and measuring functions at each position shall meet the following requirements:

1 For the cooling water supply system, the drain mains for turbine-generator unit and main transformer shall be each equipped with a flow component, which shall be able to monitor the flow at this point and output analog or digital signals.

2 The pump outlet pipe should be equipped with a flow component, which outputs switch quantity signals, both switch quantity and analog signals, or both switch quantity and digital signals.

3 The cooler outlet pipe and fire main shall be each equipped with a flow component, which outputs switch quantity signals, both switch quantity and analog signals, or both switch quantity and digital signals.

4 The water supply pipes for main shaft seal and upper and lower labyrinth rings shall be each equipped with a flow component, which outputs switch quantity signals, both switch quantity and analog signals, or both switch quantity and digital signals.

5 The water supply pipes for equipment lubrication shall be each equipped with a flow component, which outputs switch quantity signals, both switch quantity and analog signals, or both switch quantity and digital signals.

6 The oil pipes for the external circulation systems shall be each equipped with a flow component, which outputs the switch quantity signals, both switch quantity and analog signals, or both switch quantity and digital signals.

3.0.2 The alarm and shutdown values of flow monitoring and measuring should meet the following requirements:

1 The low flow alarm value for each cooler outlet pipe should be determined and provided by the equipment manufacturer, and should not be less than 70 % of the design flow rate if not specified by the manufacturer.

2 The low flow alarm value for the pump outlet pipe should not be less than 70 % of the design flow rate.

3 For bearing lubrication pipe of tubular unit, the low flow alarm and

shutdown value should be determined and provided by the equipment manufacturer. If not specified by the manufacturer, the low flow alarm value should not be less than 70 % of the design flow rate and the shutdown value should not be less than 50 % of the design flow rate.

4　The low flow alarm value and shutdown value for water supply pipe of the main shaft seal should be determined and provided by the equipment manufacturer. If not specified by the manufacturer, the low flow alarm value should not be less than 70 % of the design flow rate, and the shutdown value should not be less than 50 % of the design flow rate.

3.0.3 The switch quantity, analog or digital signal output from the flow components shall meet the data acquisition requirements of the computer supervisory control system.

3.0.4 The long-term operating environmental conditions of flow components shall comply with the current sector standard DL/T 1107, *Basic Specifications of Automatic Components for Hydropower Plant*.

4 Selection of Monitoring Points

4.0.1 The arrangement of flow monitoring points for unit auxiliary system should be in accordance with Table 4.0.1.

Table 4.0.1 Arrangement of flow monitoring points for unit auxiliary system

No.	Position	Number of monitoring points	Flow component
1	Drain main for unit cooling water system	1	Electromagnetic flowmeter
2	Water supply pipe for main shaft seal	1	Flowmeter
3	Drain main for air coolers	1	Flow switch
4	Water outlet pipe for thrust bearing cooler	1	Flow switch
5	Water outlet pipe for upper guide bearing cooler	1	Flow switch
6	Water outlet pipe for lower guide bearing cooler	1	Flow switch
7	Water outlet pipe for turbine guide bearing cooler	1	Flow switch
8	Cooling water supply pipe for upper labyrinth ring	1	Flow switch
9	Cooling water supply pipe for lower labyrinth ring	1	Flow switch
10	Lubrication oil inlet pipe for generator guide bearing of tubular unit	1	Mechanical flowmeter
11	Lubrication oil inlet pipe for turbine guide bearing of tubular unit	1	Mechanical flowmeter
12	Lubrication oil inlet pipe for two-direction thrust bearing of tubular unit	1	Mechanical flowmeter
13	Oil outlet pipe of external circulation cooler for thrust bearing, turbine guide bearing, etc.	1 each	Mechanical flowmeter

4.0.2 The arrangement of flow monitoring points for public auxiliary system should be in accordance with Table 4.0.2.

Table 4.0.2 Arrangement of flow monitoring points for public auxiliary system

No.	Position	Number of monitoring points	Flow component
1	Outlet pipe of cooling water supply pump, fire pump, etc.	1/set	Flow switch
2	Lubricating water supply pipe of deep-well pump for dewatering system	1/set	Flow switch
3	Outlet pipe of dewatering pump	1/set	Flow switch
4	Lubricating water supply pipe of deep-well pump for leakage drainage system	1/set	Flow switch
5	Outlet pipe of leakage drainage pump in the underground powerhouse	1/set	Flowmeter
6	Drain main for main transformer coolers	1	Electromagnetic flowmeter
7	Branch water drain pipe for main transformer coolers	1/set	Flow switch
8	Oil circulation main pipe for main transformer coolers	1	Mechanical flowmeter
9	Chilled water outlet pipe of air conditioning chiller	1	Flow switch
10	Cooling water outlet pipe of air conditioning chiller	1	Flow switch
11	Water outlet pipe for static frequency converter (SFC) cooler	1	Flow switch
12	Water outlet pipe for air compressor cooler	1/set	Flow switch
13	Water outlet pipe of oil sump cooler for pressure oil unit	1	Flow switch

5 Selection of Flow Components

5.0.1 The flow components for oil system pipes should use mechanical flowmeters with a sight glass. The flowmeter used for the oil pipe of external circulation system of main transformer shall be explosion-proof.

5.0.2 The flow components shall meet the following basic technical requirements:

1 The flow components shall be reliable in operation, stable in performance, and simple in structure, and shall be easy to install, test and maintain.

2 Alarm signal shall be sent out when the flow rate in the pipe reaches the set value, with a permissible error of ±5 %.

3 The flow alarm set value should be adjustable within a range of 0.25 to 1.00 times the measuring range.

4 The analog output should be 4 mA to 20 mA current signal.

5.0.3 The electromagnetic flowmeter shall meet the following requirements:

1 The flowmeter shall have the local data display function to read the instantaneous flow rate and accumulative flow volume.

2 The flowmeter shall be able to output 4 mA to 20 mA analog signal and two channels of switch quantity signals simultaneously.

3 The measurement accuracy should not be lower than Grade 2.5.

4 The flow velocity measurement range shall be 0 m/s to 8 m/s.

5 The measured pipe diameter shall be DN25 to DN1000.

6 The power supply shall be 24 VDC or 220 VAC.

5.0.4 The mechanical flowmeter shall meet the following requirements:

1 The head loss shall be less than 0.1 m.

2 The flowmeter should be provided with service flanges, which is easy to remove and clean. The flowmeter with a sight glass may be selected.

3 The flowmeter shall be able to output 4 mA to 20 mA analog signal and two channels of switch quantity signals simultaneously.

4 The switch may reset when there is motionless pressure fluid in the pipe or when the pipe is empty.

5 The measurement accuracy should not be lower than Grade 2.5.

6   The flow velocity measurement range shall be 0 m/s to 8 m/s.

7   The power supply shall be 24 VDC or 220 VAC.

5.0.5   The thermal conductivity flow switch shall meet the following requirements:

1   The flow measuring range and alarm value shall be adjustable.

2   The meter shall have anti-scaling function.

3   The switch shall reset when there is motionless pressure fluid in the pipe or when the pipe is empty.

4   The flow velocity measurement range shall be 0 m/s to 3 m/s.

5   The response time shall be 2 s to 10 s.

6   The power supply shall be 24 VDC or 220 VAC.

6 Layout and Installation Requirements of Pipes

6.0.1 The flow components should be mounted in such positions as to facilitate observation and operation.

6.0.2 The installation position of flow components should avoid strong magnetic field or strong vibration.

6.0.3 For the pipes requiring accurate flow measurement, the flow component should be mounted to a straight pipe section, and the section length before and after the flow component should meet the following requirements:

1. For an electromagnetic flowmeter, the length of the straight section before the flowmeter should not be less than 5 times the pipe diameter, and the length after should not be less than 3 times the pipe diameter.

2. For a mechanical flowmeter, the length of the straight section before the flowmeter should not be less than 5 times the pipe diameter, and the length after should not be less than 3 times the pipe diameter.

3. For a flow switch, the length of the straight section before the flow switch should not be less than 3 times the pipe diameter, and the length after should not be less than 2 times the pipe diameter.

6.0.4 When a plug-in electromagnetic flowmeter is used for DN25 to DN80 pipe, a special tube socket may be adopted. For DN100 or larger, a pipe fitting may be used.

6.0.5 For vertical pipes, the flow components should be installed on the pipe with up flow.

6.0.6 For horizontal pipes, the plug-in flow components should be mounted on the side of the pipe within 45° to the horizontal axis, where the insertion depth of electromagnetic flowmeter probes shall be 0.125 to 0.25 times the pipe diameter, and the probe of thermal conductivity flow switch shall protrude from the inner wall of the pipe.

Explanation of Wording in This Specification

1. Words used for different degrees of strictness are explained as follows in order to mark the differences in executing the requirements in this specification.

    1) Words denoting a very strict or mandatory requirement:

        "Must" is used for affirmation; "must not" for negation.

    2) Words denoting a strict requirement under normal conditions:

        "Shall" is used for affirmation; "shall not" for negation.

    3) Words denoting a permission of a slight choice or an indication of the most suitable choice when conditions permit:

        "Should" is used for affirmation; "should not" for negation.

    4) "May" is used to express the option available, sometimes with the conditional permit.

2. "Shall meet the requirements of…" or "shall comply with…" is used in this specification to indicate that it is necessary to comply with the requirements stipulated in other relative standards and codes.

List of Quoted Standards

DL/T 1107, *Basic Specifications of Automatic Components for Hydropower Plant*